HOMELAND SECURITY
OPERATIONAL ANALYSIS CENTER

Characterizing the Performance of Uncrewed Aircraft Systems

BRADLEY WILSON, SHANE TIERNEY, RACHEL M. BURNS

This research was published in 2023.

Approved for public release; distribution is unlimited.

About This Report

To become more informed about the capabilities of uncrewed aircraft systems (UASs) and to support modeling and simulation capabilities,[1] the U.S. Department of Homeland Security (DHS) Science and Technology Directorate (S&T) asked the Homeland Security Operational Analysis Center (HSOAC) to help it characterize the performance of UASs of all sizes. The resulting analysis, in which we documented UAS platforms' speed, payload capacity, and endurance, will inform S&T's modeling and simulation and other activities and help it provide relevant feedback to DHS component stakeholders. A better understanding of UAS capabilities will also enhance research, development, and acquisition activities across DHS. The intended audience for this report is DHS components and interagency stakeholders that are interested in UAS performance.

This research was sponsored by the S&T Office of Science and Engineering and conducted within the Acquisition and Development Program of the HSOAC federally funded research and development center (FFRDC).

About the Homeland Security Operational Analysis Center

The Homeland Security Act of 2002 (Section 305 of Public Law 107-296, as codified at 6 U.S.C. § 185) authorizes the Secretary of Homeland Security, acting through the Under Secretary for Science and Technology, to establish one or more FFRDCs to provide independent analysis of homeland security issues. The RAND Corporation operates HSOAC as an FFRDC for DHS under contract HSHQDC-16-D-00007.

The HSOAC FFRDC provides the government with independent and objective analyses and advice in core areas important to the department in support of policy development, decisionmaking, alternative approaches, and new ideas on issues of significance. The HSOAC FFRDC also works with and supports other federal, state, local, tribal, and public- and private-sector organizations that make up the homeland security enterprise. The HSOAC FFRDC's research is undertaken by mutual consent with DHS and is organized as a set of discrete tasks. This report presents the results of research and analysis conducted under task order 70RSAT21FR0000085.

The results presented in this report do not necessarily reflect official DHS opinion or policy.

For more information on HSOAC, see www.rand.org/hsoac. For more information on this publication, see www.rand.org/t/RRA1566-2.

---

[1] In the past, the most common term in this field was *unmanned*. We opt for the more inclusive *uncrewed*.

Acknowledgments

We thank the sponsoring staff in S&T for their guidance and support. We also thank Andrew Karode and Angela Putney of HSOAC for their reviews and Lauren Skrabala and Lisa Bernard for editing the report.

Summary

Issue

To gain a clearer picture of uncrewed aircraft system (UAS) capabilities to support modeling and simulation efforts, the U.S. Department of Homeland Security (DHS) Science and Technology Directorate (S&T) asked the Homeland Security Operational Analysis Center to help it characterize the performance and availability of various UAS platforms. The resulting analysis expands on an effort for S&T that explored the performance of small-UAS platforms by characterizing the performance of UASs of *all sizes* and investigating a subset of agricultural UASs. The resulting analysis, in which we documented UAS platform speed, payload capacity, and endurance, will inform S&T's modeling and simulation and other activities and help it provide relevant feedback and risk-informed decision support to DHS component and interagency stakeholders. A better understanding of UAS capabilities will also enhance research, development, and acquisition activities across DHS.

Approach

Our analysis leveraged available data sets that aggregated UAS performance, starting with a data set that we used in previous work for DHS, data on air platforms from the Association for Uncrewed Vehicle Systems International's Uncrewed Systems and Robotics Database. To build a more comprehensive set of performance data, we identified and integrated other data sources where possible, including the Center for a New American Security's Drone Database, part of the center's Proliferated Drones project; the Norwegian University of Science and Technology's Unmanned Aerial Vehicles Laboratory; earlier versions of the air platform data; and information from a multitude of manufacturer resources, such as websites, brochures, and social media posts.

After assembling the data, we performed a validation and verification exercise. Validation consisted of checks for reasonableness, missingness, conversions, and cross-metric performance (speed, endurance, and weight), as well as a review of performance outliers. Verification consisted of checking the air platform data against other data sources.

We then cleaned and normalized the aggregated data set, eliminated duplicates from multiple data sets, and created several data subsets that excluded military platforms and platforms that are no longer actively sold, ensuring that the analysis focused on a set of platforms of interest to potential adversaries as noted by DHS. We considered all types, sizes, and payloads of aircraft except those with obtainability exclusions. We reported summary statistics of interest to the sponsor for the baseline data set and an agriculture-centric data set. We report the summary data in box-and-whisker plots to show the distribution of data.

Finally, we developed a novel approach to characterize spray coverage as a function of swath, speed, and endurance and presenting performance parameters of interest. The goal of spraying could be the largest area covered or it could be the spray density of the substance. We included both in the analysis.

Findings

We discovered gaps, but no major issues, in the Association for Uncrewed Vehicle Systems International air platform data from its Uncrewed Systems and Robotics Database. For example, price data were missing for 86 percent of platforms, a handful of fields had misaligned unit conversions (e.g., miles per hour to kilometers per hour), and 37 percent of status fields (representing the current availability of platforms) were out of date. However, these shortfalls did not affect the results of the analysis or the strength of our conclusions. We were able to correct 28 percent of the values for the performance parameters in our analysis, but, importantly, the number of those changes was not statistically significant. Furthermore, upon examining the performance outliers, we found them all to be justified. Therefore, we conclude that, although the data are not perfect, they appear to be a good representation of manufacturer-reported performance.

We modeled a combination of speed, endurance, payload, spray rate, and spray height to explore the platforms' spray coverage, as shown in Figure S.1.

FIGURE S.1

Spray Coverage in the Baseline Data Set, by Platform Size

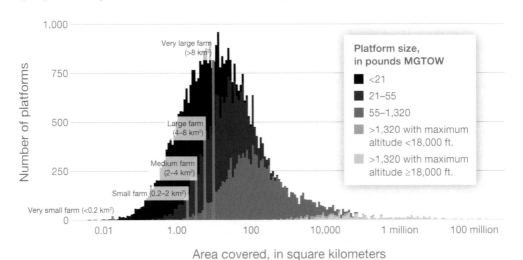

SOURCE: Farm sizes are based on MacDonald, Korb, and Hoppe, 2013.

NOTE: km² = square kilometers. MGTOW = maximum gross takeoff weight. The horizontal axis uses a log base 10 scale. Some aircraft with MGTOW greater than 1,320 lb. and a maximum altitude of at least 18,000 ft. (shown in yellow) specialize in high-altitude flight, and their endurance and speed assumptions might not hold in lower atmospheric maneuvers.

We found that reasonable assumptions about spray rate (based on pump capability) and height were far more–important drivers of spray coverage than any of the other UAS performance measures:

- Changing the pump rate from 1.8 kg per minute to 8 kg per minute resulted in a 77-percent average decrease in the potential area covered. We saw a similarly large trend when we increased the spray height but did not observe statistically significant change in potential area covered with changes in endurance, which indicates that payload and dispersal rate, rather than aircraft endurance, were the drivers of area coverage.
- Changing the spray height 650 percent from 10 ft. to 75 ft. resulted in an equal 650-percent average increase in the potential area covered. This is because the geometry of the spray is assumed to be constant independent of height.
- Increasing cruise speed by 1 km per hour resulted in a 3-percent average increase in the potential area covered.
- Increasing payload weight by 1 kg was associated with an average increase of 17 percent in the potential area covered.

Endurance (i.e., the amount of time that a platform can operate) and the endurance scenario indicator (i.e., whether the model used 100 or 50 percent of the endurance value) were not statistically significant predictors of potential area covered after we controlled for the other factors in our analysis. Platform endurance is not a major factor in these models because the total time spraying is almost always limited by pump rate and payload weight rather than endurance.

Finally, we assessed the density of aerosolization and found that smaller UAS platforms tend to provide denser coverage (0.001 kg per square meter) than larger platforms (in the baseline and agriculture data set) do, likely because of their slower overall cruise and maximum speeds. Also, changing spray rate and spray height naturally changes the spray density. Determining significant spray densities of specific farming chemicals was outside the scope of this work. In this project, we did not attempt to optimize the usage of larger platforms. It remains to be seen how much platforms can be slowed below their cruise speeds under load to maximize saturation.

Next Steps

This research could serve as the foundation for several interesting avenues of exploration for S&T in support of modeling and simulation capability development. For example, it could support the development of additional data sets of price information, inform a detailed analysis of third-party test data, and serve as a starting point for data-set verification that employs a larger sample. Other next steps could include developing a parameterized analysis to look more deeply at the trade space between use case, platform, and coverage and building component-level models of selected platforms.

Contents

Figures and Tables

Figures

Tables

CHAPTER ONE

Issue and Data Development

Issue

The U.S. Department of Homeland Security (DHS) Science and Technology Directorate (S&T) asked the Homeland Security Operational Analysis Center to leverage previous work (Wilson et al., 2020) for S&T that characterized the performance of small uncrewed aircraft systems (sUASs). The analysis presented in this report expanded on that effort by characterizing performance of all sizes of uncrewed aircraft systems (UASs) (not just small platforms) and by investigating agricultural UAS performance due to UASs' growth in the agricultural market segment. Analysis of UAS speed, payload capacity, endurance, and spraying capability will inform S&T's modeling and simulation and help S&T provide relevant feedback and risk-informed decision support to stakeholders from components across DHS and partners across the homeland security enterprise. A better understanding of UAS capabilities will also enhance research, development, and acquisition activities across DHS.

In the remainder of this chapter, we describe the development of the data used to characterize the performance. Chapter Two characterizes the performance data, and Chapter Three presents conclusions and next steps. A supplementary data table generated at the request of S&T can be found in the appendix.

Performance Parameters

To inform parameters for modeling UASs, S&T was interested in UAS platforms' ability to carry useful payloads, reach target locations, and spread potentially aerosolized payloads on those locations. We focused on speed, payload capacity, maximum gross takeoff weight (MGTOW), and endurance. We default to the most–commonly reported units for each (i.e., knots for speed, pounds for payloads and takeoff weights). We discuss each of these values in this section.

Speed
Speed determines a platform's ability both to reach distant targets and cover large areas when deploying its payload. Aircraft most often operate at cruise speed, a designed operating speed

that is set based on the aircraft's role. We factored both cruise speed and maximum speed into our parametric analysis of spray coverage in Chapter Two.

Payload Capacity and Maximum Gross Takeoff Weight

Payload capacity is determined by an aircraft's size, power, and configuration. In this way, it is closely linked to MGTOW. Larger and more-powerful aircraft can carry larger payloads. Aircraft configuration can also affect payload capacity. For instance, fixed-wing platforms can often carry larger payloads than their rotary-wing counterparts, but they might require additional infrastructure, such as landing strips, to do so. Fixed-wing aircraft also lack the low-speed maneuverability of rotary-wing aircraft. We use U.S. Department of Defense UAS categories (Joint Chiefs of Staff, undated, Chapter Two).

Endurance

Endurance determines how long an aircraft can stay airborne. Adding payload or operating at speeds far from the speed for maximum endurance (often designed to be near cruise speed) can reduce this value. For our analysis, we also had to consider another form of endurance: that of the payload. Once the payload tank is emptied, the aircraft can no longer disperse the substance and is no longer useful. We determined this time parametrically in our analysis, using values derived from commercial pumps designed for agricultural UASs. In our analysis, we examined both types of endurance to determine the period when the platforms were able to actively support agricultural operations.

Spray Coverage

All of these parameters factor into a more agriculture-specific metric, spraying capability, which we call *spray coverage*. We examined how long a UAS can provide spray coverage based on its endurance, payload, and the area it can cover in that time, a factor of its speed and altitude. We were able to determine the total area covered while dispensing payload in a single sortie. Further discussion of this analysis and its results are presented in Chapter Two.

It is important to note that, given the volume and variety of UASs, we did not dynamically compute endurance as a function of the payload. This would require a more rigorous modeling of aircraft performance that was beyond the scope of this effort.

Approach

To accomplish the goals of this project, we leveraged available data sets that aggregated UAS performance, starting with a data set that we used in previous work for DHS (see "Data Sources," next). Next, we identified other relevant data sources that we could integrate with the baseline data set to build a more comprehensive set of performance data.

After compiling, cleaning, and normalizing the data (see "Data Assembling and Cleaning" later in this chapter), we performed a validation and verification exercise (see "Validation and Verification" later in this chapter). Once we were satisfied with the outcome of the exercise, we proceeded with the performance characterization described in Chapter Two.

We designed the characterization to explore UAS performance on the parameters of interest to the sponsor and developed a novel approach to capturing spray coverage.

Data Sources

Our main source of data was the Association for Uncrewed Vehicle Systems International's (AUVSI's) Uncrewed Systems and Robotics Database—specifically, the data on air platforms (AUVSI, undated). This large, comprehensive database offers a wealth of information on platform specifications and capabilities. We were unable to identify any other database as comprehensive as AUVSI's catalog of approximately 4,000 platforms.

For thoroughness and to augment gaps in the AUVSI data, we explored several other databases and collections of information on UASs:

- The Center for a New American Security (CNAS) Proliferated Drones project is an effort to "examine the implications of drone proliferation and identify the core issues facing the United States and its partners" (CNAS, undated a). Part of that effort was the creation of a database of proliferated UASs (CNAS, undated b). This collection of specifications for 154 air platforms is much smaller than the AUVSI data set but was useful for validating and augmenting information therein.
- Richard Hann and Joachim Wallisch at the Norwegian University of Science and Technology created a data set of 93 air platforms as part of a larger effort related to wind tunnel design (Hann and Wallisch, 2020). As a result, the database focuses on properties related to the platforms' aerodynamic performance. As with the CNAS database, we used Hann and Wallisch's database to validate and augment the AUVSI data.
- As part of our process to validate the AUVSI data, we gathered information from a multitude of manufacturer resources, including websites, brochures, and social media posts. The appendix contains a list of manufacturer sources considered.
- Janes Markets Forecast data include information on hundreds of air platforms, although they focus on military and military-adjacent platforms (Janes Markets Forecast, undated). As a result, those data had less utility for our analysis because independent actors cannot easily acquire those types of platforms. Incorporating Janes UAS data into our analysis was outside of scope of our current analysis, but they could be considered as a future avenue of exploration.
- Finally, our previous sUAS study for DHS (Wilson et al., 2020) used an earlier version of the AUVSI air platform data that we augmented, validated, and enhanced with manufacturer-provided information. We carried these additions over into our current analytic data set.

Data Assembling and Cleaning

The original 2021 AUVSI data consisted of 3,929 observations (air platforms) and 126 variables.[1] We identified 40 air platforms that were not included in the original AUVSI data set, and we added these to our analytic data set. Information about these additional platforms came from a variety of sources. We cross-referenced platforms listed in both the CNAS Drone Database and the Norwegian University of Science and Technology's Unmanned Aerial Vehicles (UAV) Laboratory against the AUVSI data and added any that were not already included in our data set.

Over the course of our research, we conducted open internet searches and discovered additional platforms listed on company websites. Some of these were new platforms from known companies, such as Da-Jiang Innovations' (DJI's) latest agricultural platforms, and others were from start-up firms. As we explored the data and conducted validation and verification exercises (see the section "Validation and Verification" later in this chapter), we created a data set with modifications to provide updated or corrected values. We wrote a program in the R statistical programming application to update the current and future versions of the AUVSI data (provided that the structure of the data remains the same). However, if a future version of the AUVSI data incorporates updates, corrections, or new values, the program will not replace those values.

Baseline Data

The purpose of this analysis was to assess the potential performance of commonly available UAS platforms with a focus on those with agricultural use cases. Consequently, we limited our analytic data set to platforms that would be accessible to civilians. As a proxy for this, we excluded platforms that the data indicated were marketed only to the military.

The AUVSI data also indicated that several platforms were marketed to both military and commercial consumer bases; we reviewed each of these platforms and opted to exclude platforms that were unlikely to be available to civilians. We used our judgment to identify

- any platform designed specifically for a military or defense purpose
- any platform from a company that had a policy or sales structure making that UAS available only as a service and not for purchase.

---

[1] The full list of variables is available in the AUVSI database (AUVSI, undated).

We also reviewed platforms that were marketed to "civil" or "civil and academic" consumers and excluded additional platforms that were unlikely to be available to civilians. We used our judgment to identify

- any platform that was a prototype
- any platform that requires contact information-gathering that could provide a logistical hurdle for acquisition
- any platform with government involvement in its development that restricted sales.

The market categorization from AUVSI is useful to an extent but is not a substitute for what we were trying to determine—the availability of the platform—which is why we reviewed these individually.

Finally, we excluded platforms that were categorized as "inactive" status in the AUVSI data. These modifications resulted in an analytic baseline data set of 1,863 platforms of interest. Figure 1.1 displays and quantifies the modifications that we made to the original data set that resulted in our final analytic data set.

FIGURE 1.1

Modifications to the Association for Uncrewed Vehicle Systems International Data Set to Create a Data Set of Platforms of Interest

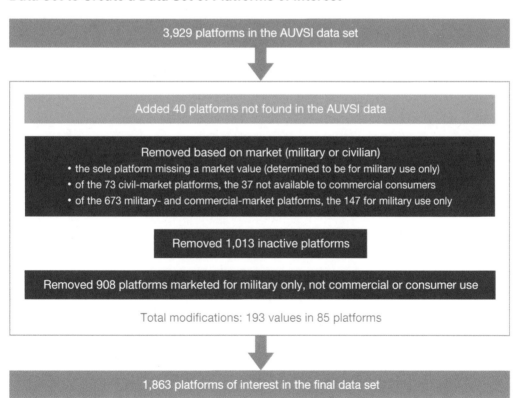

Validation and Verification

To assess the completeness and veracity of the original AUVSI data, we conducted several validation and verification exercises, which led to the modifications described in Figure 1.1. Validation consisted of checks of reasonableness, missingness, conversions, and cross-metric performance, as well as a review of performance outliers.[2] Verification consisted of checking the AUVSI data against other data sources, including manufacturing data and other third-party information, as available. We limited verification and some of the validation exercises to the platforms of interest for this project; however, we present data for the complete AUVSI data set when relevant and practical.

Validation

The goal of the validation exercises was to answer the following questions:

- Are the data reasonable (i.e., are the data in line with what we would expect)?
- Are they sufficient to address our research questions?

Missingness

We first assessed missingness in our variables of interest. These included

- a description of the platform
- status
- affiliated country or countries
- airframe
- application
- market
- energy source
- price
- several performance metrics.

Rates of missingness are displayed in Figure 1.2. Most descriptive qualitative variables had few or no missing values. For nearly all platforms, data were available on airframe type, and less than 10 percent were missing data on energy source. The data were less complete for performance metric variables, with maximum speed and cruise speed having the least-complete data (46 percent and 66 percent missing, respectively). However, only 35 percent of platforms were missing both maximum-speed and cruise-speed data; therefore, the original data pro-

---

[2] *Reasonableness* is defined as the state of being within an expected range, value, or type (i.e., whether the data are what would be expected). *Missingness* is defined as the state of being missing (i.e., whether the data are missing or not).

FIGURE 1.2

Percentages of Data Missing from the Association for Uncrewed Vehicle Systems International Data Set for Variables of Interest

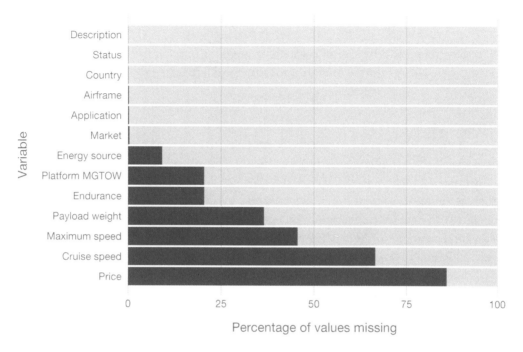

vided insights into the speed of roughly two-thirds of the platforms. Price data were missing for 86 percent of platforms and, thus, were not sufficient for a comprehensive cost analysis.

Conversions

The AUVSI data present performance metrics for more than one unit; for example, maximum speed is provided in miles per hour, kilometers per hour, and knots. We checked the conversions for speed, endurance, MGTOW, and payload weight to ensure that the reported metrics were consistent across units. We identified fewer than ten observations in each category that were significantly different across units, most likely because of conversion mistakes or data-entry errors.

Payload Anomalies

We identified 24 platforms with values for MGTOW that were less than or equal to payload weight. Of these, we identified ten platforms of interest. In this set of ten, we corrected three anomalies, removed the six platforms that were no longer active, and removed the one platform for which we were unable to determine the correct value.

Extreme Outliers

To assess data reasonableness, we identified extreme outliers in our performance metric variables of interest: speed, endurance, MGTOW, and payload weight. We defined *extreme outlier* as any value that met either of these conditions, by platform type:

- greater than the 75th percentile plus three times the interquartile range
- below the 25th percentile minus three times the interquartile range.

Figure 1.3 shows the number of outliers for each category of interest.

Of those platforms with nonmissing values, we identified extreme outliers for endurance (2.2 percent), speed (1.6 percent), payload weight (5.5 percent), and MGTOW (5.2 percent).

We identified the outliers in active platforms of interest and closely examined each value to determine whether it was anomalous and incorrect. We found that the performance metric outliers were not unusual if other factors were considered. For example, battery-powered rotorcraft with endurance outliers used hydrogen fuel cells. Speed outliers, especially for fixed-wing aircraft with internal combustion, were not unusual or unprecedented. Outliers for MGTOW and payload weight were found in larger platforms.

FIGURE 1.3

Extreme Outliers in the Association for Uncrewed Vehicle Systems International Data for Variables of Interest

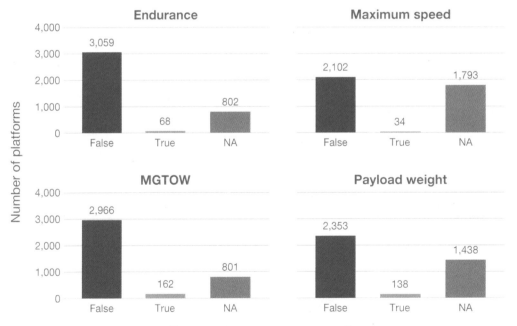

NOTE: NA = not available.

Verification

The goal of the verification exercises was to answer the question, "Are the data accurate?" The gold standard for answering this question would involve procuring UAS platforms and performing field tests; however, doing so was not practical for this project. As an alternative, we compared performance data in the AUVSI data set with data from the manufacturers' websites and third-party sources. We pulled a random sample of 5 percent of the active platforms of interest (n = 96) and compared data on market, status, MGTOW, payload weight, maximum speed, and endurance.[3] We categorized each observation as shown in Table 1.1.

Market and Status

Because platform availability is a key aspect of our analysis, we verified both the market (civilian, military, or both) and status (whether a platform is active or being actively marketed) variables to determine whether platforms would be easily obtainable. Figure 1.4 shows the results of our verification exercise.

We were able to verify most observations in our sample. We updated only six values for the market variable but 37 values for status. Most of these were platforms that AUVSI indicated were actively marketed; however, these data might be out of date.

Performance Metrics

We also verified a sample of platforms' performance metrics—specifically, MGTOW, payload weight, maximum speed, and endurance. We updated the MGTOW values for 14 platforms (15 percent), payload weight values for ten platforms (10 percent), maximum-speed values for five platforms (5 percent), and endurance values for ten platforms (10 percent). We were able to fill in missing data for MGTOW (one platform), payload weight (one platform), and maximum speed (five platforms). Figure 1.5 shows the full results of the verification exercise for the performance metrics.

We compared the original AUVSI data with the values that we updated during the verification exercise and plotted the percentage change to assess the impact of the updates. We found that only a few platforms had large shifts in values, as shown in Figure 1.6.

TABLE 1.1

Categories of Data in Our Verification Process

Category	Description
Verified	The AUVSI data were consistent with other sources.
Updated	We changed the original AUVSI value or replaced a missing AUVSI value.
Unable to verify	No manufacturer or third-party data source was available for verification.
Missing	The AUVSI data did not have a value for that field and we were unable to fill in the missing data.

---

[3] We cannot determine the sufficiency of this sample size without verifying more of the data.

FIGURE 1.4

Findings from the Verification of Market and Status Variables

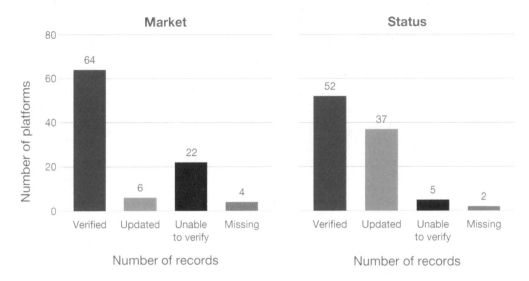

To confirm that these changes did not affect the trends, we conducted *t*-tests to compare the pre- and postverification means but did not find a statistically significant difference in means for any of the four performance metrics. Figure 1.7 shows the means and distribution of the performance metrics both before and after verification.

In Chapter Two, we characterize the performance of the baseline data set.

FIGURE 1.5

Findings from the Verification of Performance Metrics

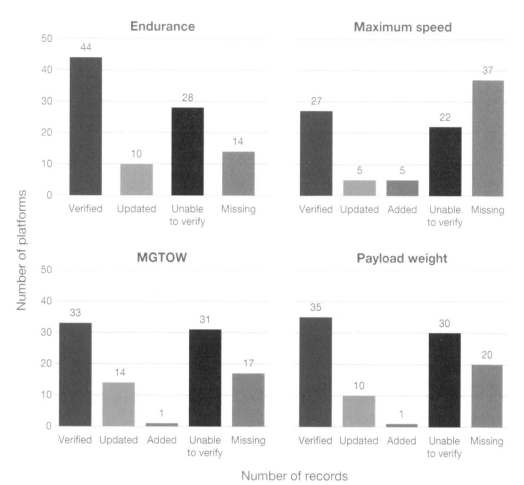

NOTE: Endurance does not include any additions.

FIGURE 1.6

Percentage Changes in Performance Metric Values After Verification

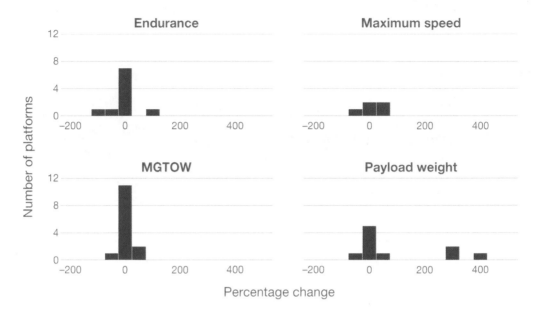

FIGURE 1.7

Distribution of Performance Metrics in the Verified Sample Data

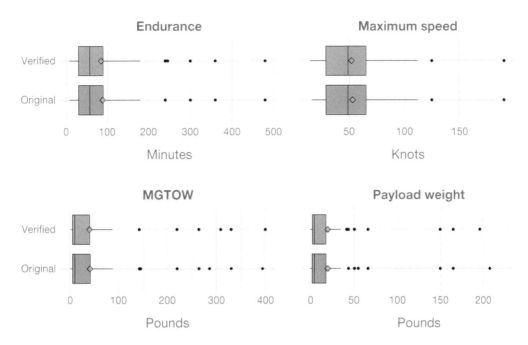

NOTE: Our *t*-tests showed no statistically significant difference in means between the original sample of AUVSI values and the verified sample.

CHAPTER TWO

Performance Characterizations

In this chapter, we first characterize the performance parameters for the baseline data set of interest. We then present an additional subset of the baseline, an agricultural application data set, to represent sets of platforms specifically suited for agricultural spraying use cases.

The Baseline Analysis

This section summarizes the characteristics and performance metrics of the baseline data-set platforms of interest. S&T stakeholders might be focused on numerous capabilities, so this set serves to help them respond to potentially broad questions.

Platform Type

We first categorized data based on airframe and energy source. Rotary-wing platforms powered by a battery are the most common type of platform in the baseline data set, followed by fixed-wing battery and fixed-wing internal combustion platforms (see Figure 2.1).

Status

Most of the platforms in our baseline data set were listed as active or actively marketed (74.5 percent). Figure 2.2 shows the distribution of the status variable and the frequency and percentage of other status categories.

Maximum Gross Takeoff Weight

More than half of the platforms in the baseline data set were small platforms with an MGTOW of less than 21 lb. Figure 2.3 shows the distribution of platforms by MGTOW.

Maximum Speed

The average maximum speed for platforms in the baseline data set is 54.3 knots; the median is 48 knots. Figure 2.4 shows the distribution of maximum speed by platform type.

FIGURE 2.1

Platforms in the Baseline Data Set, by Airframe and Energy Source Type

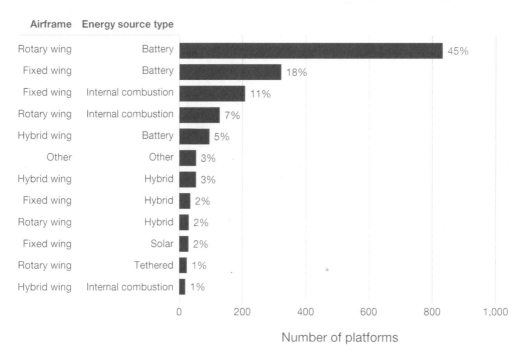

Number of platforms

FIGURE 2.2

Platforms in the Baseline Data Set, by Status

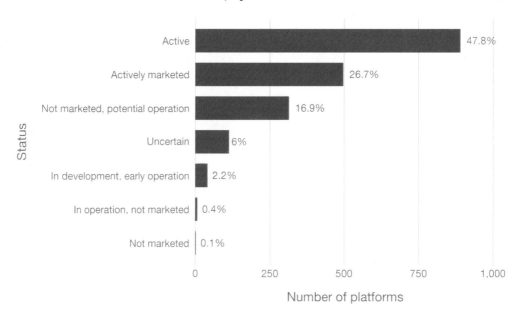

Number of platforms

FIGURE 2.3

Platforms in the Baseline Data Set, by Department of Defense Uncrewed Aircraft System Maximum Gross Takeoff Weight Class

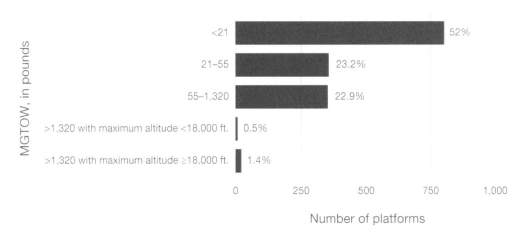

Number of platforms

FIGURE 2.4

Distribution of Maximum Speed, by Platform Type, in the Baseline Data Set

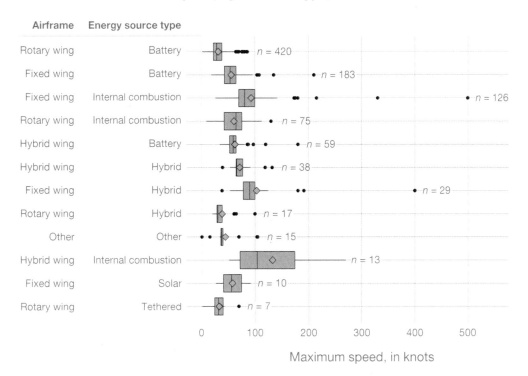

Maximum speed, in knots

Endurance

The median endurance for all platforms in the baseline data set is 60 minutes. There are several extreme outliers in the endurance data, so the median endurance value is likely more meaningful than the mean endurance value, which is 109 hours. Figure 2.5 shows the distribution of endurance by platform type.

Maximum Payload Weight

The median payload weight for all platforms is 10.6 lb.; the mean is 289 lb. Figure 2.6 shows the distribution of payload weight by platform type.

Spray Coverage and Density

We needed to create a metric to determine the relative effectiveness of UAS agricultural spraying capability. We chose to measure this by examining the area that each platform could cover in a single sortie under the specified performance metrics (speed, endurance, and pay-

FIGURE 2.5

Distribution of Endurance, by Platform Type, in the Baseline Data Set

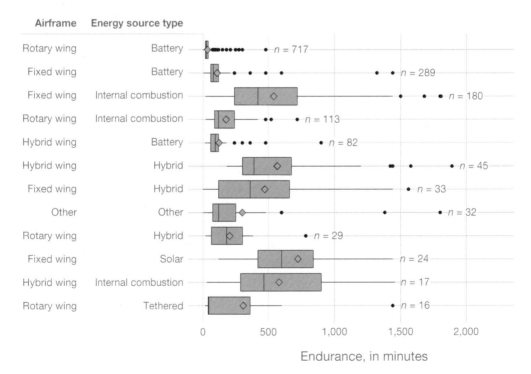

NOTE: The figure excludes 41 observations with endurance values of more than 2,000 minutes. These excluded values range from 40 to 43,800 hours. These were mainly solar-powered, tethered, or hybrid platforms.

FIGURE 2.6

Distribution of Payload Weight, by Platform Type, in the Baseline Data Set

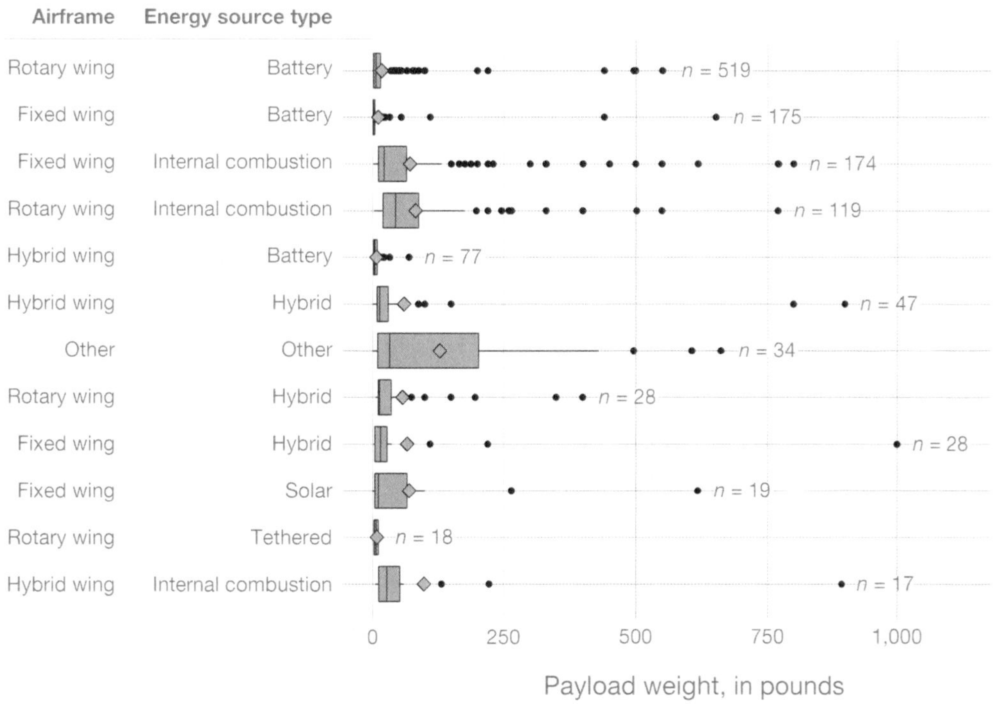

Airframe	Energy source type	
Rotary wing	Battery	n = 519
Fixed wing	Battery	n = 175
Fixed wing	Internal combustion	n = 174
Rotary wing	Internal combustion	n = 119
Hybrid wing	Battery	n = 77
Hybrid wing	Hybrid	n = 47
Other	Other	n = 34
Rotary wing	Hybrid	n = 28
Fixed wing	Hybrid	n = 28
Fixed wing	Solar	n = 19
Rotary wing	Tethered	n = 18
Hybrid wing	Internal combustion	n = 17

Payload weight, in pounds

NOTE: There are 19 platforms not shown on the figure that have payload weight values ranging from 1,014 to 200,000 lb. Most of these are large, fixed-wing platforms that rely on internal combustion.

load weight). To examine the absolute effectiveness of these platforms, we also explored the density of coverage using water as a proxy for the payload.

To determine the spray coverage, we measured the area covered under typical factors for agricultural spraying use cases, as a function of spray swath, speed, and endurance. The equation for the area covered by a sortie is as follows: area covered = swath × speed × functional endurance. In the rest of this section, we discuss each of these factors.

Speed

Because of the high rates of missingness for the cruise-speed variable, we imputed cruise-speed values when cruise speed was missing but maximum-speed data were available. For platforms that had both cruise speed and maximum speed, we calculated the ratio of cruise speed to maximum speed and the 25th and 75th percentiles of that ratio. To obtain two imputed cruise-speed variables, we then multiplied those values by the maximum-speed value when the cruise speed was missing and the maximum speed was nonmissing.

Endurance

We also needed to calculate how long the platform would be over the intended area. Two ends of a spectrum were selected to represent the variation in use. First, we imagined a scenario in which the platform was launched immediately adjacent to the intended area, spending 100 percent of its time over the area. Then we imagined a scenario in which the platform had to be launched from far away and use half its flight endurance reaching the intended area. Further examination of endurance suggests that there are two factors to consider:

- the stated flight endurance of the platform, which depends on the fuel
- the payload endurance, or how long the platform can spray out its payload before running out.

This creates a logical equation for what we call *functional endurance*:

functional endurance = (flight endurance, in minutes × the number of minutes over the intended area) or (maximum payload weight, in pounds ÷ pump rate, in kilograms per minute), whichever is lower.

Swath

To calculate swath, we first considered that these platforms have a typical height of approximately 10 ft. above the intended area. Using a survey of agricultural spraying platforms in the database and attempting to fit a linear relationship between the maximum payload weight and spray swath results in the following swath width:[1] swath = (0.2 × maximum payload weight) + 3.6. This gives the swath for typical operation, but a user might use a platform in unusual ways not captured here. To account for this, we parameterized the height of release at the typical height of 10 ft. over the intended area, as well as heights of 25, 50, and 75 ft.[2] Note that this will also result in decreased particle density on the target from these greater heights: The payload delivery rate does not change to accommodate the greater area covered. We assumed, for simplicity, that the payload dispersal would follow roughly the same pattern as it does in typical operation.[3] We could then calculate that the swath from other heights by realizing that they produce similar triangles, as shown in Figure 2.7.

---

[1] The survey consisted of 15 currently commercially available platforms. The coefficient of determination (R^2) value of the linear fit is 0.9243.

[2] We ignored potential environmental hazards to flight when considering these parameters and included all cases in the current analysis. Under real-world conditions, a fast-flying aircraft, for example, would likely avoid flying as low as 10 ft. above the terrain because of the risk of hitting ground obstacles, such as trees and walls.

[3] In reality, droplet dispersal is likely to vary from this result because of environmental conditions (e.g., wind) and the reduction in rotor downwash (i.e., the change in surrounding airflow) farther from the vehicle. Computing spray swath with these considerations in mind would become a complex fluid dynamics problem that was outside the scope of this analysis.

FIGURE 2.7

Spray Swath at Varying Heights

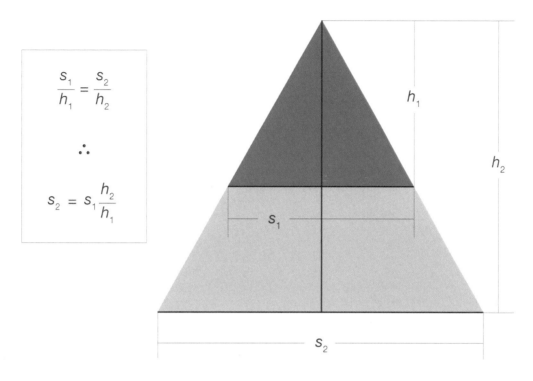

NOTE: s = swath. h = height. $\therefore$ = therefore.

A survey of commercial pumps for agricultural platforms did not indicate a consistent relationship between pump speed and payload or MGTOW. As a result, we chose to parameterize this value as well. We set these values to align with those of pumps for commercial platforms, using 1.8, 3, and 8 kg per minute as the low, medium, and high values. It is important to note that these calculations do not take environmental conditions into account. So such considerations as wind and temperature do not affect the results.

In total, we had 48 combinations of factors, or scenarios, from a full factorial combination of four height parameters, three pump-rate parameters, two endurance parameters, and two cruise-speed parameters. We were able to obtain spray area covered for 905 (49 percent) of our platforms of interest, resulting in a data set of 43,392 observations. Table A.1 in the appendix shows the mean and median potential spray coverage for available AUVSI platforms, where spray time is limited only by endurance, speed, and payload (i.e., no unforeseen issues deter spraying or flight). There are some larger platforms with above-average speed and endurance that are skewing the data to the right; therefore, the median values for each scenario might be a better indicator of average potential spray area for each of the scenarios. We present the area-covered data on a log base 10 scale for improved visualization and ease of interpretation. Figure 2.8 shows that the log-transformed data on area covered are still somewhat skewed to

FIGURE 2.8

Log-Transformed Area Covered Across the Full Factorial of Scenarios

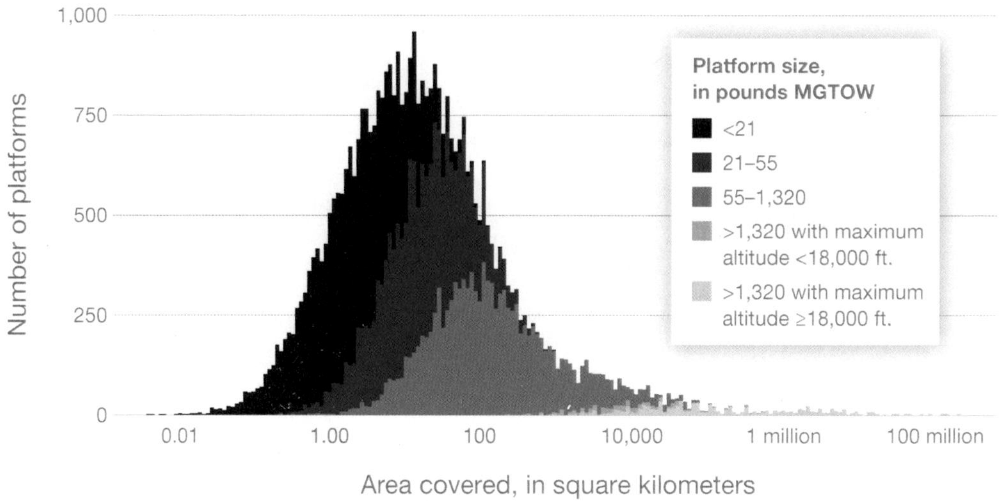

NOTE: The MGTOW classification is marked not available, but all the performance data necessary to calculate spray coverage exist. The horizontal axis uses a log base 10 scale.

the right after transformation; most of these data points are large platforms with high cruise speeds and large payload weight capacity.

To clarify the relative influence of each of the parameters used in our calculations for area covered, we generated a linear regression model that predicted the log-transformed area covered in square kilometers with the parameters from the calculations. Table 2.1 shows the model output.

Changing the pump rate from 1.8 kg per minute to 8 kg per minute resulted in a 77-percent average decrease in potential area covered, meaning that endurance can generally be understood to be limited by the rate at which payload is delivered rather than by aircraft endurance. Changing the spray height from 10 ft. to 75 ft. resulted in a 650-percent average increase in potential area covered due to the assumed constant geometry of the spray pattern at all heights, although we again note that real-world factors, such as wind and drifting particles, make this assumption less reliable as height increases. Thus, changing the spray rate and height produced significant change to spray coverage roughly equivalent to the change in unit. Changing the spray rate from 1.8 kg per minute to 8 kg per minute yielded a 77-percent increase in rate that showed a corresponding 77-percent decrease in spray coverage because the platform is spraying more quickly.

Other significant effects included increasing cruise speed by 1 km per hour, resulting in a 3-percent average increase in potential area covered; using the 75th-percentile ratio of cruise speed to maximum speed for imputing missing cruise speed resulted in an average increase in potential area covered of 18 percent. Increasing payload weight by 1 kg was associated with an average increase in potential area covered of 17 percent. Endurance and the endurance

TABLE 2.1

Output from an Ordinary-Least-Squares Linear Regression Model Predicting Log-Transformed Area Covered, in Square Kilometers, with Parameters from Calculation

Parameter	Estimate	Standard Error	t-Statistic	p-Value
(Intercept)	0.26021	0.028335	9.183373	0.00000
Cruise speed, in kilometers per hour, imputed at 25th-percentile ratio of maximum speed to cruise speed	0.03242	0.000194	166.8272	0.00000
Cruise-speed imputation at 75th percentile versus the 25th percentile	0.16415	0.018024	9.107662	0.00000
Endurance, in hours	0.00002	2.19E-05	0.78459	0.43270
Endurance scenario: 50% versus 100%	−0.00884	0.018024	−0.49049	0.62379
Payload weight, in kilograms	0.00168	2.6E-05	64.60621	0.00000
Spray rate, in kilograms per minute, versus 1.8 kg per minute				
3	−0.50180	0.022075	−22.7322	0.00000
8	−1.47424	0.022075	-66.7848	0.00000
Spray height, in feet, versus 10 ft.				
25	0.91629	0.025489	35.94784	0.00000
50	1.60944	0.025489	63.14132	0.00000
75	2.01490	0.025489	79.04849	0.00000

scenario indicator were not statistically significant predictors of potential area covered after we controlled for the other factors in the model.

Platform endurance is not a major factor in these models because the total time spraying is almost always limited by the pump rate and payload weight rather than by endurance, even in cases in which half the aircraft's endurance time is spent in transit to and from the target (see Figure 2.9).

Table 2.2 shows how the potential spray area increases with increasing spray height and decreasing pump rate.

Table 2.3 shows the distribution of potential area covered, by platform type. On average, internal combustion engines can cover more area than battery. These observations follow previous observations on internal combustion and fixed-wing platforms tending toward greater speed.

Figure 2.10 shows the strong relationship between platform size, in MGTOW, and the potential area covered. Note that the x-axis is presented on a log 10 scale.

Although some larger platforms can cover vast areas, the area covered has an inverse relationship with saturation (see Figure 2.11), and, assuming that the payload is evenly dispersed

FIGURE 2.9

Total Time-Limiting Factor for Platforms in the Baseline Analysis, by Pump Rate

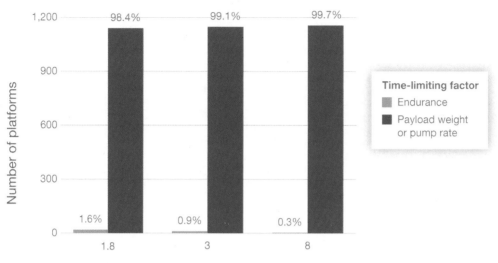

TABLE 2.2

Potential Area Covered Across the Full Factorial of Scenarios, by Spray Height and Pump Rate

Scenario	Mean	Standard Deviation	Minimum	Q1	Median	Q3	Maximum
Spray height, in feet							
10	24,582	631,933	0.00	1.1	4.7	22	27,456,741
25	61,455	1,579,832	0.01	2.7	11.8	56	68,641,852
50	122,910	3,159,665	0.02	5.4	23.7	112	137,283,704
75	184,365	4,739,497	0.03	8.1	35.5	168	205,925,556
Pump rate, in kilograms per minute							
1.8	158,171	4,331,282	0.02	6.4	28.1	136	205,925,556
3	98,955	2,606,258	0.01	3.8	16.9	82	123,555,334
8	37,857	978,170	0.00	1.4	6.3	31	46,333,250

NOTE: Q1 = first quartile. Q3 = third quartile.

over the area covered, the density or saturation levels might be too low to have the desired effect, depending on what substance is used.

In this section, we described the methods and assumptions we used to understand area that sprayers can cover and density or saturation levels; however, running these models on the full baseline data set can give misleading results. In the next section, we describe the per-

TABLE 2.3

Potential Area Covered Across the Full Factorial of Scenarios, by Platform Type

Airframe	Energy Source Type	n	Mean	Standard Deviation	Minimum	Q1	Median	Q3	Maximum
Fixed wing	Battery	150	380	3,954	0.00	1.2	3.5	11	113,542
	Hybrid	27	334,815	1,751,050	0.41	12.8	58.6	383	22,917,531
	Internal combustion	147	84,219	684,459	0.43	25.5	92.8	480	14,515,875
	Solar	10	352	914	0.09	3.0	15.6	123	5,985
Hybrid wing	Battery	61	25	55	0.07	2.5	7.5	22	619
	Hybrid	37	4,540	35,347	0.35	14.3	39.5	112	503,892
	Internal combustion	14	4,780,427	23,198,063	4.37	57.7	184.6	2,020	205,925,556
Rotary wing	Battery	326	64	475	0.02	1.6	6.5	25	14,070
	Hybrid	17	558	2,260	0.73	8.2	20.9	54	25,171
	Internal combustion	85	2,345	8,138	0.20	35.0	139.0	876	111,667
	Tethered	6	19	30	0.14	2.4	7.1	23	173
Other		13	2,138	8,724	0.00	4.6	37.4	323	74,994

formance metrics, spray coverage, and spray density of platforms that already have sprayer capabilities (our agricultural application data set). We excluded platforms that would not be practical or capable of carrying out agricultural spraying scenarios (e.g., high-altitude solar gliders).

FIGURE 2.10

Potential Area Covered Across the Full Factorial of Scenarios, by Department of Defense Unmanned Aircraft System Maximum Gross Takeoff Weight Class

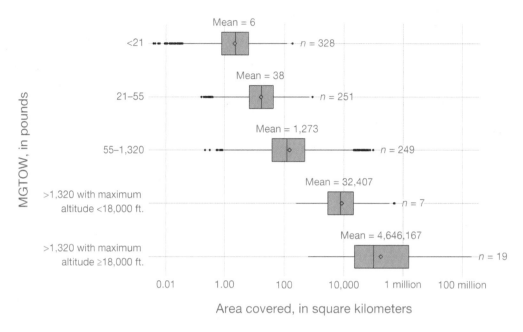

NOTE: The horizontal axis uses a log base 10 scale.

The Agricultural Application Data Set

Platform Type

To facilitate understanding of the characteristics and capabilities of platforms with agricultural use cases, we created an additional subset of the data for analysis from the baseline data set. We developed the agricultural application data set to focus on platforms that were intended for agricultural applications, such as crop sprayers. This data set includes platforms from the baseline data set with an agricultural application and excluded platforms that also listed "intelligence, surveillance, reconnaissance" or "imaging" (the Application field allows multiple values). We also excluded two tethered platforms from the data (one rotary-wing platform and one ornithopter). This data set is significantly smaller than the baseline data set, containing only 66 platforms, as shown in Figure 2.12.

FIGURE 2.11

Density of Spray for Maximum Area Covered

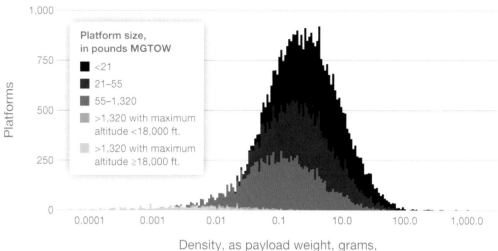

NOTE: The MGTOW classification is marked not available, but all the performance data necessary to calculate spray coverage exist. The horizontal axis uses a log base 10 scale.

FIGURE 2.12

Platforms in the Agricultural Application Data Set, by Airframe and Energy Source Type

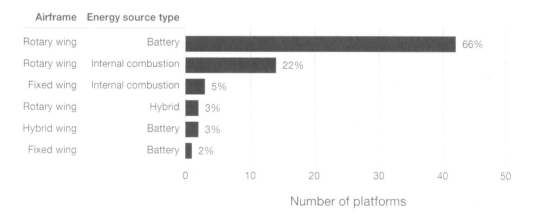

Status

The platforms in this set are largely in active use and actively marketed, as shown in Figure 2.13.

FIGURE 2.13

Platforms in the Agricultural Application Data Set, by Status

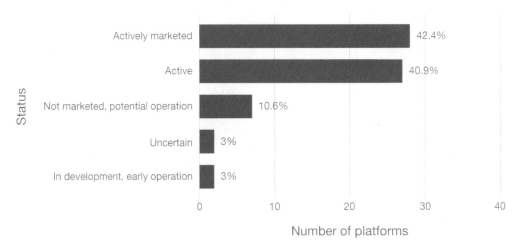

Maximum Gross Takeoff Weight

This set contains mostly small and medium-sized platforms, as shown in Figure 2.14. This is unsurprising, given the results of our analysis described in the previous section and the fact that their targets are crop fields. Furthermore, platforms with MGTOW less than 55 lb. require no special permission or licensing to use, driving demand for this type among cost-conscious commercial users.

Maximum Speed

Maximum-speed trends follow those observed in the baseline data set with regard to platform type and power source. Figure 2.15 shows that most platforms in this set are battery-powered rotary-wing aircraft.

FIGURE 2.14

Platforms in the Agricultural Application Data Set, by Maximum Gross Takeoff Weight

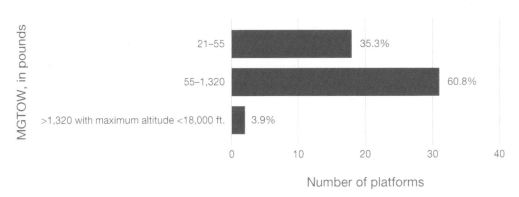

FIGURE 2.15

Distribution of Maximum Speed, by Platform Type, in the Agricultural Application Data Set

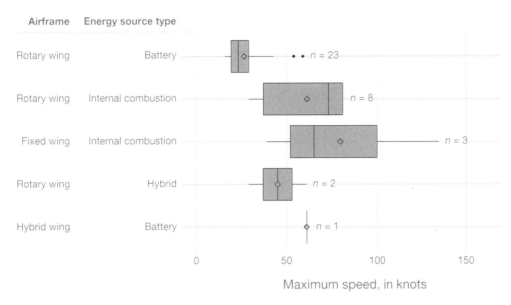

NOTE: Because some data were missing, not all combinations of airframe and energy source type are represented in the data set.

Endurance

In Figure 2.16, we see results that mirror those of the larger data set: Battery-powered rotary-wing aircraft often have shorter endurance than their internal combustion and fixed-wing counterparts have.

FIGURE 2.16

Distribution of Endurance, by Platform Type, in the Agricultural Application Data Set

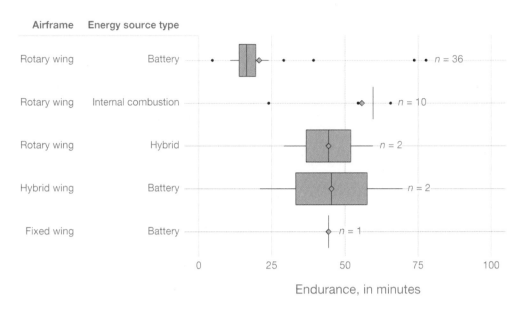

NOTE: The figure excludes one rotary-wing internal combustion platform with an endurance value of 480 minutes. Because some data were missing, not all combinations of airframe and energy source type are represented in the data set.

Maximum Payload Weight

Figure 2.17 shows that internal combustion and fixed-wing options might present a more attractive option for hauling larger payloads. Of course, these come with greater operations and maintenance demands and greater takeoff and landing requirements, respectively.

FIGURE 2.17

Distribution of Payload Weight, by Platform Type, in the Agricultural Application Data Set

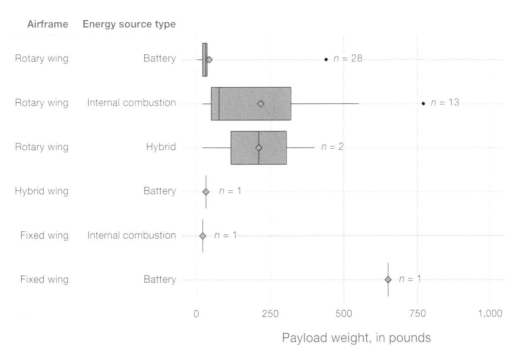

NOTE: The figure excludes one rotary-wing internal combustion platform with a payload weight of 1,380 lb. Because some data were missing, not all combinations of airframe and energy source type are represented in the data set.

Spray Coverage and Density

We were able to calculate spray area covered for 30 platforms (45 percent) in the set, resulting in a data set of 1,440 observations (accounting for four height parameters, three pump-rate parameters, two endurance parameters, and two cruise-speed parameters).

Figure 2.18 shows that the time-limiting factor for the platform is usually a function of the platform emptying its payload before reaching maximum endurance.

When we used a pump rate of 1.8 kg per minute, endurance, rather than payload weight or pump rate, was the limiting factor on total time spraying in 15 percent of the platforms in the set. The percentage of platforms limited by endurance decreased to 10 percent and 3 percent for scenarios using pump rates of 3 kg per minute and 8 kg per minute, respectively.

In Figure 2.19, we see the spray coverage for the scenarios on a log base 10 scale.

Table 2.4 shows the range of performance results by airframe and energy source type, with the one fixed-wing platform performing the best overall.

Figure 2.20 shows the area covered using the MGTOW categories, demonstrating the area covered growing as the category increases. Note that the x-axis is shown on a log 10 scale.

FIGURE 2.18

Total Time-Limiting Factor for Platforms in the Agricultural Application Data Set, by Pump Rate

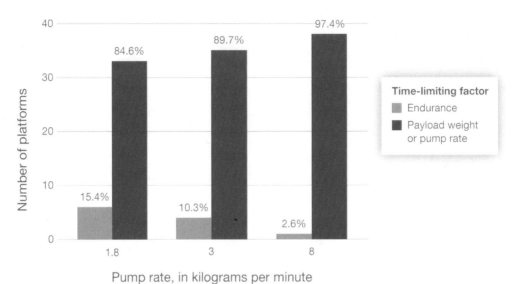

FIGURE 2.19

Potential Area Covered Across the Full Factorial of Scenarios, Agricultural Application Data-Set Platforms

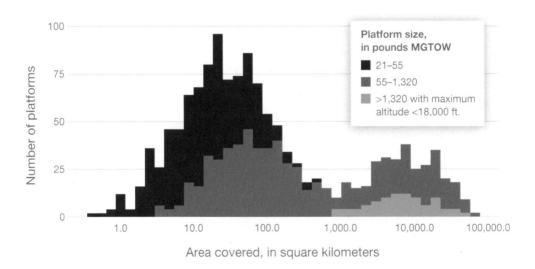

NOTE: The MGTOW classification is marked not available, but all the performance data necessary to calculate spray coverage exist. There is an order-of-magnitude increase in area covered at each higher MGTOW category. The horizontal axis uses a log base 10 scale.

TABLE 2.4

Potential Area Covered Across the Full Factorial of Scenarios, by Platform Type, Agricultural Application Data-Set Platforms

Airframe	Energy Source Type	n	Mean	Standard Deviation	Minimum	Q1	Median	Q3	Maximum
Fixed wing	Battery	1	19,536	14,257	3,392	8,056	16,961	26,030	50,882
Hybrid wing	Battery	1	201	177	19	74	131	278	619
Rotary wing	Battery	18	60	110	0	9	23	64	1,120
	Hybrid	2	3,559	5,490	3	29	509	5,759	25,171
	Internal combustion	8	11,194	17,187	3	275	4,157	14,171	108,338

NOTE: Because some data were missing, not all combinations of airframe and energy source type are represented in the data set.

FIGURE 2.20

Potential Area Covered Across the Full Factorial of Scenarios, by Maximum Gross Takeoff Weight Class, Agricultural Application Data-Set Platforms

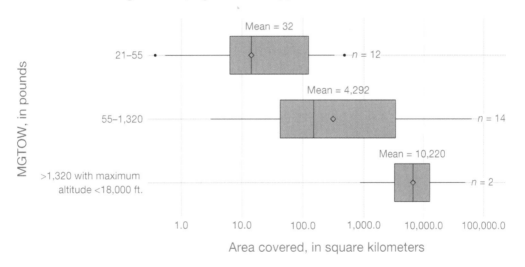

NOTE: Because some data were missing or the data set was too small, not all MGTOW categories are represented in the data set. The horizontal axis uses a log base 10 scale.

Finally, in Figure 2.21, we combine data on the potential area covered and payload weight to calculate the density of aerosolization.

This metric reflects the saturation of spray at the platform's potential area covered. The necessary density to achieve desired farming effects will differ depending on the agent used (e.g., more or less diluted fertilizer). However, we can see from this analysis that the larger

FIGURE 2.21

Density for Agricultural Application Data-Set Platforms

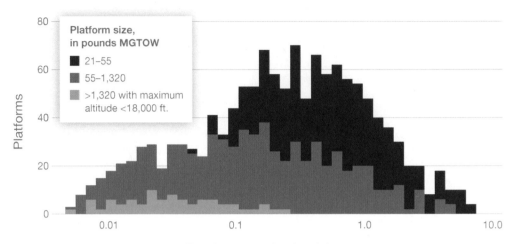

NOTE: The MGTOW classification is marked not available, but all the performance data necessary to calculate spray coverage exist. The horizontal axis uses a log base 10 scale.

platforms tend to move more quickly. Even though they carry an increased payload, their rate of saturation is lower than that of the smaller platforms.

Chapter Three presents our conclusions and next steps to build on these analyses.

Conclusions and Next Steps

Conclusions

We can draw several conclusions from this characterization of UAS performance data. Our verification and validation exercises showed that the AUVSI data were largely consistent with data from other sources and that performance outliers were reasonable. We were able to characterize the performance of the platforms in the data set and subsets of interest using the requested parameters. To explore spray coverage, we generated parameters for potential area covered and density, which are functions of several existing parameters.

We found that the data on platform price were insufficient for characterization of platform accessibility. Furthermore, the infrequency of platform obsolescence updates limits our understanding of the current availability of platforms.

This analysis was a first step in providing insight into UAS performance. Aircraft performance is a trade space, and models of performance could be developed to provide more analytic rigor.

The rest of this section describes our conclusions in more detail.

Data Development Limitations

Following the validation and verification exercises, we identified several limitations to the AUVSI data.

Limited Price Data

The price data were inadequate for a comprehensive cost analysis, which would have provided additional insights into platform accessibility. The AUVSI data were missing 86 percent of prices. However, this information might be available from other sources or compiled from manufacturer and retail websites.

Unit Conversions

We found very few conversion errors or data anomalies; most of these were likely data-entry errors, and we corrected them in our analysis.

Obsolete Status Data

We updated 37 percent of the status-field data (indicating whether platforms were actively marketed); for most platforms, we could not confirm whether they were still being actively marketed. This indicates that the AUVSI data have not been adequately updated to reflect the ever-changing marketplace.

Data Development Strengths

We also identified several positive elements from the validation and verification exercise.

Performance Outliers

Our inspection of outliers indicated that platforms with performance metric data on the long tails of the distributions were not mistakes but legitimate values (e.g., platforms with extremely long endurance were often solar-powered gliders capable of flying for long periods).

Verifiable Performance Data

After completing the verification analysis of a random sample of the data, we determined that most of the data (85 to 95 percent) from the sample were accurate, and the updates that we found would not affect an analysis of average performance metrics for available platforms.

Overall, we concluded that the data accurately represent manufacturer-reported data. Although it is important to note that, because of the labor intensity of the verification exercise, we explored a sample of only 5 percent of the data set. The changes made during the verification produced a negligible change in the overall performance characterization.

Characterization

Forty-five percent of the baseline platforms of interest are rotary-wing battery-electric (the most common), compared with 66 percent in the agricultural use data set.

Status

In both analytic data sets, most platforms were active or actively marketed, indicating that they should be obtainable to potential users.

Maximum Gross Takeoff Weight

MGTOW is an area in which the agricultural set differs substantially from the baseline. Whereas, in the baseline, 52 percent of platforms have an MGTOW less than 55 lb. (i.e., sUAS), the agricultural use data set has no platforms in that category indicating that, although sUASs are the majority of the UASs available, spraying use cases are not exclusively the domain of sUASs.

Maximum Speed

Despite a few outliers, maximum speeds tend to cover a range from 50 to 100 knots. The agricultural data-set platforms tended to be slower than average, likely because a smaller share of them were fixed-wing platforms.

Endurance

Battery-electric UASs have significantly less endurance than other types of platforms. Battery-powered platforms (save for a few outliers) do not endure beyond four hours, and we could expect even less under load. On the other hand, in the baseline set, there are five classes of platforms (airframe and energy source) not solely reliant on batteries with a maximum endurance of close to 24 hours.

Only one platform in the agricultural use set had an endurance beyond 80 minutes, and nearly all had endurance less than one hour.

Maximum Payload Weight

This parameter is more difficult to characterize. We found rotary-wing platforms performing better overall, particularly in the agricultural use set.

Spray Coverage

Finally, we modeled a combination of speed, endurance, payload, spray rate, and spray height. We found that reasonable assumptions about spray rate (based on pump capability) and height were more-impactful drivers of spray coverage than any of the UAS performance measures.

Changing the pump rate from 1.8 kg per minute to 8 kg per minute resulted in a 77-percent average decrease in potential area covered, indicating that payload rather than platform endurance is the limiting factor in most scenarios. Changing the spray height from 10 ft. to 75 ft. resulted in a 650-percent average increase in potential area covered because of the assumed constant geometry of the spray pattern at all heights. Increasing cruise speed by 1 km per hour resulted in a 3-percent average increase in potential area covered. Increasing payload weight by 1 kg was associated with an average increase in potential area covered of 17 percent. Endurance and the endurance scenario indicator were not statistically significant predictors of potential area covered when we controlled for the other factors in the model. Platform endurance is not a major factor in these models because the total time spraying is almost always limited by pump rate and payload weight as opposed to endurance.

Every platform in the agricultural use set was able to cover an area of at least 3 km².

We also assessed the density of the aerosolization and found that sUASs tended to have better density (0.001 kg per square meter) than larger platforms (in both the baseline and the agricultural use set), likely because of sUASs' slower overall cruise and maximum speeds. In this work, we did not attempt to optimize usage of larger platforms. Future work could address how much these platforms can be slowed below their cruise speeds under load to maximize saturation.

Next Steps

Several potential next steps present themselves as interesting avenues of exploration to further inform this analysis:

- An effort could be made to cross-reference the database values with Janes air platform data to incorporate that collection's more-robust price data. Although Janes analysis focuses on military platforms, there might be enough overlap to make this a worthwhile endeavor.
- Incorporating sufficient price data from Janes or other sources would allow further filtering of the analysis data sets for how obtainable a platform is based on cost tiers.
- Access to third-party test data would strengthen the verification process. Our verification consisted of testing AUVSI's stated values against those from other databases and manufacturer-provided sources. Third-party platform testing would reveal how trustworthy these provided data are. If sufficient analysis cannot be found, S&T might seek to conduct this testing itself.
- Increasing the size of the verification sample could build further confidence in the data set.
- Our general, parameterized analysis could be applied to a scenario analysis that focuses on areas of concern for DHS. Such an analysis could incorporate greater specificity into combinations of parameterized values, including limiting them to those that create meaningful densities of distributed payload.
- Future work could explore the trade space between use case, platform, coverage, and saturation level in greater detail.

Supplementary Data

In Table A.1, we summarize the spray coverage by scenario.

Table A.2 lists manufacturer and third-party sources used to verify and validate AUVSI values. Note that the total number of platforms listed here does not add up to 5 percent of the total count. This is because some manufacturers have shuttered, sales of their platforms have ceased, and no other records of platform performance exist. In these cases, the only validation possible is validating the market status as "inactive" and the only "source" is a lack of sources. Because there are so many start-ups and the UAS industry changes so quickly, particularly for sUASs, this is a somewhat common result.

TABLE A.1
Statistical Summary of Potential Spray Coverage, by Scenario

Scenario Parameter				Area Covered, in Square Kilometers			
Spray Height, in Meters	Cruise-Speed Imputation, as a Percentile	Endurance Scenario	Spray Rate, in Kilograms per Minute	Median	Mean	Minimum	Maximum
10	25	Best	1.8	8	40,450	0.02	27,456,741
	75	Best	1.8	11	41,186	0.02	27,456,741
	25	Best	3.0	5	24,930	0.01	16,474,044
	75	Best	3.0	6	25,376	0.01	16,474,044
	25	Best	8.0	2	9,384	0.00	6,177,767
	75	Best	8.0	2	9,556	0.01	6,177,767
	25	Worst	1.8	8	38,021	0.02	27,456,741
	75	Worst	1.8	11	38,514	0.02	27,456,741
	25	Worst	3.0	5	24,105	0.01	16,474,044
	75	Worst	3.0	6	24,545	0.01	16,474,044
	25	Worst	8.0	2	9,374	0.00	6,177,767
	75	Worst	8.0	2	9,542	0.01	6,177,767
25	25	Best	1.8	21	101,126	0.04	68,641,852
	75	Best	1.8	26	102,966	0.06	68,641,852
	25	Best	3.0	13	62,325	0.02	41,185,111
	75	Best	3.0	16	63,440	0.04	41,185,111
	25	Best	8.0	5	23,461	0.01	15,444,417

Table A.1—Continued

Scenario Parameter				Area Covered, in Square Kilometers			
Spray Height, in Meters	Cruise-Speed Imputation, as a Percentile	Endurance Scenario	Spray Rate, in Kilograms per Minute	Median	Mean	Minimum	Maximum
	75	Best	8.0	6	23,890	0.01	15,444,417
	25	Worst	1.8	21	95,053	0.04	68,641,852
	75	Worst	1.8	26	96,286	0.06	68,641,852
	25	Worst	3.0	13	60,263	0.02	41,185,111
	75	Worst	3.0	16	61,363	0.04	41,185,111
	25	Worst	8.0	5	23,435	0.01	15,444,417
	75	Worst	8.0	6	23,856	0.01	15,444,417
50	25	Best	1.8	42	202,252	0.08	137,283,704
	75	Best	1.8	53	205,931	0.12	137,283,704
	25	Best	3.0	25	124,650	0.05	82,370,222
	75	Best	3.0	32	126,879	0.07	82,370,222
	25	Best	8.0	10	46,922	0.02	30,888,833
	75	Best	8.0	12	47,780	0.03	30,888,833
	25	Worst	1.8	42	190,107	0.08	137,283,704
	75	Worst	1.8	53	192,572	0.12	137,283,704
	25	Worst	3.0	25	120,525	0.05	82,370,222
	75	Worst	3.0	32	122,727	0.07	82,370,222
	25	Worst	8.0	10	46,870	0.02	30,888,833

Table A.1—Continued

Scenario Parameter				Area Covered, in Square Kilometers			
Spray Height, in Meters	Cruise-Speed Imputation, as a Percentile	Endurance Scenario	Spray Rate, in Kilograms per Minute	Median	Mean	Minimum	Maximum
75	75	Worst	8.0	12	47,712	0.03	30,888,833
	25	Best	1.8	63	303,379	0.12	205,925,556
	75	Best	1.8	79	308,897	0.18	205,925,556
	25	Best	3.0	38	186,974	0.07	123,555,334
	75	Best	3.0	47	190,319	0.11	123,555,334
	25	Best	8.0	14	70,383	0.03	46,333,250
	75	Best	8.0	18	71,670	0.04	46,333,250
	25	Worst	1.8	63	285,160	0.12	205,925,556
	75	Worst	1.8	79	288,857	0.18	205,925,556
	25	Worst	3.0	38	180,788	0.07	123,555,334
	75	Worst	3.0	47	184,090	0.11	123,555,334
	25	Worst	8.0	14	70,305	0.03	46,333,250
	75	Worst	8.0	18	71,568	0.04	46,333,250

TABLE A.2

List of Manufacturer and Third-Party Sources

Platform	Source Considered
500 X4	CNAS, undated b
AeroDrone MR4	Bask Aerospace, undated b
AeroDrone MR6	Bask Aerospace, undated a
Aeromapper EV2	Aeromao, undated b; ntc_tech, 2021
Aeromapper Talon	Aeromao, undated a
AgEagle	AgEagle Aerial Systems, undated
Agras T10	DJI, undated a
Agras T30	DJI, undated b
Airboxer	HighEye, undated
Alpha 800	Alpha Unmanned Systems, undated
AR100-C	AirRobot, undated
ASN-209	China National Aero-Technology Import and Export Corporation, 2011; CNAS, undated b; Mjaawad, 2011
Berkut 2E	CNAS, undated b
Bixler 3	HobbyKing, undated
Blacklion 168 (BL-168)	Shandong Black Shark Intelligent Technology, undated
Cabure III	Airforce Technology, 2011; CNAS, undated b
CamCopter SL	Airstarintl, undated
Cardinal II	National Chung-Shan Institute of Science and Technology, 2020
Carrier H6 Hybrid HE+	Harris Aerial, undated
CineTank MK 2	FlyingCinema, undated
Clairvoyance	CNAS, undated b
Cobra	Thunder Tiger, undated
CPXD-6R-16L	Challenger Precision Multi-Copters, 2019
Creator	BORMATEC, undated
Curiosity	OM UAV Systems, undated
CX-180 ThunderHawk	Thunder Tiger, undated
Da Vinci	Flying Production, undated
Dauntless	Mobile Recon Systems, undated
Delta Y	Delta Drone, undated; Delta Drone, 2017

Table A.2—Continued

Platform	Source Considered
DP-12 Rhino	Dragonfly Pictures, undated
E-7	CNAS, undated b; Edaran Mesra, undated
e.Yo 200	e.Yo, undated
EHang 216 (Logistics)	EHang, undated
Ehecatl	CNAS, undated b; La Franchi, 2007
Elektra Two Solar	Elektra Solar, undated
Evo II	Autel Robotics, undated
Flying EYE	360 Designs, undated
Goshawk W200	Lozano Rocabeyera, 2016
HARRIER	Challenger Aerospace Systems, 2019a
Harrier Industrial	VulcanUAV, undated
HE190E	Helipse, undated
HEF 80	HighEye, undated
Helicam	Helicam, undated
Hexo+	Hexo+, undated
KC-09	"AI Bird KC-09," 2017; "AI Bird UAV," 2020
KUS-FT	CNAS, undated b
KWT-Z4M-C	Shenzhen Keweitai Enterprise Development, undated
KWT-Z4M-H	Shenzhen Keweitai Enterprise Development, undated
LEAP	Unmanned Systems and Solutions, undated
Logo 500 SE	Mikado Model Helicopters, undated
LP500	CNAS, undated b
Mercury-1	CNAS, undated b; Parmar, 2015
Minion	Hoffer, 2014
Navig8 Electric	4Front Robotics, undated
Nomad B	CNAS, undated b
NT4Contras	AirVision, undated
Octopush	Repreneurs.com, 2021
Oxygen	Shenzhen Joyton Innovation Technology, undated
P40	XAG, undated a

Table A.2—Continued

Platform	Source Considered
Patroller	CNAS, undated b, Safran, undated
PD8X	Prodrone, undated
PLT05	Beijing Pilotless UAV Technology, undated; "PUT UAV," 2020
PR-5 Wiewiór	"PR-5 Wiewior Plus," 2022; Samoloty w Lotnictwie Polskim, undated; "SKNL PRz PR-5 Wiewior" [Технические характеристики. Фото Подробнее на], 2016
Q01	CNAS, undated b; "Reiner Stemme Q01," 2020
Q4E	Mavtech, undated
R-16	Shenzhen Xiangnong Innovation Technology, undated
Recon	UASUSA, undated
REMO M-002	UCONSYSTEM, undated a; UCONSYSTEM, undated b; UCONSYSTEM, undated c
Rheinmetall KZO	CNAS, undated b
Rotor Buzz	UAV Research Lab, undated
Rover XS	Integrated Dynamics, 2019a
Scythe	microUAV, undated
Sentinel G1	Aero Sentinel, undated
Siemari OS-190	Aerofoundry, undated
Sirius I	Topcon, 2015
Sisuar	CNAS, undated b; INAV, undated
SkyEye Sierra VTOL	ElevonX, 2020
Sky Eye Micro	Sky Eye, undated
SKY-26D	Trancomm Technologies [版权所有], undated
SkyMapper	jDrones, undated
SkyRanger R70	CNAS, undated b; Teledyne FLIR, undated
Solaris	Integrated Dynamics, 2019b
SPARROWHAWK	Challenger Aerospace Systems, 2019b
Spider 103	Collective Wisdom Technology, undated
SR-1 Eagle Owl	SteelRock Technologies, 2019
SR-15	Rotomation, undated
Stardust	CNAS, undated b; IDETEC Unmanned Systems, undated
Surveyor-H UVH-120E	Beijing Creaton Technology, 2012; UAVOS, undated

Table A.2—Continued

Platform	Source Considered
T16	Beijing Creaton Technology, 2012
T21	Tianyu Chuangtong Technology [北京天宇创通科技有限公司 版权所有], undated
T23E Eleron-3	"T23E UAV: Eleron-3," 2012
T-80/TRV-80 (Tactical Resupply Vehicle)	Malloy Aeronautics, undated
TBS DISCOVERY PRO (Team BlackSheep)	Crashpilot, undated
Teal One	Teal Drones, undated
TILT Ranger	Inkonova, undated
TIM-X150	Chengdu Timestech, date unknown
TIM-X80	Chengdu Timestech, undated
Titan	Challenger Aerospace, 2019c
Typhoon H Plus	CNAS, undated b; Yuneec Holding, undated
UPX-23L	Beijing Proto Unmanned Aerial Vehicle Technology [北京普洛特无人飞行器科技有限公司 京], undated
UX11	Delair, undated
V40	XAG, undated b
VA002	Vanilla Unmanned, undated
Vigilant	Harpia Tech, undated
Vigilant	CNAS, undated b
Volacopter LZ-73-PRO	Volacom, undated
WINGO S	UAVision, undated
X-8	Birdpilot, undated
XAG XP 2020	Specialized Agricultural Services, undated
Xena	OnyxStar, undated
YMR-01	Yamaha Motor Corporation, undated
ZALA 421-06	Zala Aero Group, undated b
ZALA 421-16E2	Zala Aero Group, undated c
ZALA 421-21	Zala Aero Group, undated a
Zephyr II	Matthew, 2015
Zip 2	Petrova and Kolodny, 2018; "Zipline (Drone Delivery)," 2022

Abbreviations

AUVSI   Association for Uncrewed Vehicle Systems International

CNAS   Center for a New American Security

DHS   U.S. Department of Homeland Security

DJI   Da-Jiang Innovations

MGTOW   maximum gross takeoff weight

NA   not available

Q1   first quartile

Q3   third quartile

S&T   Science and Technology Directorate

sUAS   small uncrewed aircraft system

UAS   uncrewed aircraft system

UAV   uncrewed aerial vehicle

References

360 Designs, "Flying EYE™," product listing, undated. As of April 26, 2022:
https://360designs.io/product/flying-eye/

4Front Robotics, "Unmanned Aerial Vehicles (UAVs)," webpage, undated. As of May 9, 2022:
https://www.4frontrobotics.com/uavs

Aero Sentinel, "Military UAV Sentinel G1: Tactical, Foldable, Micro UAV," webpage, undated. As of April 26, 2022:
https://www.aero-sentinel.com/military-drones/military_drone_sentinel_g1/

Aerofoundry, "Drone profissional Hewë 190," AeroExpo, undated. As of April 26, 2022, in Portuguese:
https://www.aeroexpo.online/pt/prod/aerofoundry/product-181473-26652.html

Aeromao, "Aeromapper Talon," webpage, undated a. As of April 25, 2022:
https://www.aeromao.com/products/aeromapper-talon/

———, "Professional UAV Aeromapper EV2," AeroExpo, undated b. As of April 25, 2022:
https://www.aeroexpo.online/prod/aeromao/product-181549-26889.html

AgEagle Aerial Systems, "Drone Hardware," webpage, undated. As of April 25, 2022:
https://ageagle.com/drone-hardware/

"AI Bird KC-09," blog post, *Avia.pro*, August 6, 2017. As of May 7, 2022:
https://avia-pro.net/blog/ai-bird-kc-09-tehnicheskie-harakteristiki-foto

"AI Bird UAV," *Wikipedia*, Wikimedia Foundation, last edited August 2, 2020. As of April 26, 2022:
https://en.wikipedia.org/wiki/AI_Bird_UAV

Airforce Technology, "Nostromo Cabure Unmanned Aerial Vehicle (UAV)," webpage, June 19, 2011. As of May 9, 2022:
https://www.airforce-technology.com/projects/nostromo-uav/

AirRobot, untitled webpage, undated. As of April 25, 2022:
https://www.airrobot.de/en-gb/products

Airstarintl, homepage, undated.

AirVision, "NT4Contras: Il primo inimitabile controtante," webpage, undated. As of April 26, 2022, in Italian:
http://airvision.it/portfolio-items/nt4contras/

Alpha Unmanned Systems, "Alpha 800: Durability for the Long Haul," webpage, undated. As of April 25, 2022:
https://alphaunmannedsystems.com/product/uav-helicopter-alpha-800/

Association for Uncrewed Vehicle Systems International, "Unmanned Vehicle Systems and Robotics Database: Air," webpage, undated.

Autel Robotics, "Evo II," webpage, undated. As of April 26, 2022:
https://www.autelrobotics.com/productdetail/1.html

AUVSI—*See* Association for Uncrewed Vehicle Systems International.

Bask Aerospace, homepage, undated a. As of April 25, 2022:
https://baskaerospace.com.au/

————, "Introducing the All New AeroDrone MR4," webpage, undated b. As of April 25, 2022:
https://baskaerospace.com.au/aerodrone/mr4/

Beijing Creaton Technology, "T16 UAS," webpage, January 11, 2012.

Beijing Pilotless UAV Technology, webpage listing products [了解产品 > 飞机], undated.

Beijing Proto Unmanned Aerial Vehicle Technology [北京普洛特无人飞行器科技有限公司京], "UPX-23L electric intelligent plant [UPX-23L电动智能植保机]," webpage, undated. As of April 26, 2022, in Chinese:
http://www.upuav.com/site/cn/product/info/2016/2346.html

Birdpilot, "Professional UAV X-8," AeroExpo, undated. As of May 9, 2022:
https://www.aeroexpo.online/prod/birdpilot/product-181429-23017.html

BORMATEC, "Branchenlösungen" [Industry solutions], webpage, undated. As of April 25, 2022, in German:
https://www.bormatec.com/branchenloesungen

Center for a New American Security, "Proliferated Drones: About the Project," webpage, undated a.

————, "The Drone Database," webpage, undated b.

Challenger Aerospace Systems, "HARRIER UAS System," brochure, c. 2019a. As of April 26, 2022:
https://challengeraerospace.com/wp-content/uploads/2019/04/HARRIER-UAS_S.pdf

————, "SPARROWHAWK UAS System," brochure, c. 2019b. As of April 26, 2022:
https://challengeraerospace.com/wp-content/uploads/2019/04/SPARROWHAWK-UAS_S.pdf

————, "Titan," brochure, c. 2019c. As of April 26, 2022:
https://challengeraerospace.com/wp-content/uploads/2019/04/TITAN.pdf

Challenger Precision Multi-Copters, "Drone Sprayer for Agriculture," brochure, c. May 2019. As of May 9, 2022:
https://challengeraerospace.com/wp-content/uploads/2019/05/CPXD-6R-16L-Specifications.pdf

Chengdu Timestech, homepage, undated.

————, title unknown, webpage, date unknown.

China National Aero-Technology Import and Export Corporation, "ASN Series UAV System," webpage, 2011.

CNAS—See Center for a New American Security.

Collective Wisdom Technology, "Spider 103 Unmanned Aerial System," webpage, undated. As of April 26, 2022:
http://www.collective-wisdom-technology.com/en/spider/

Crashpilot, "TBS DISCOVERY PRO," FPV Blog, undated. As of May 7, 2022:
https://www.fpvblog.com/tbs-discovery-pro

"Creaton UAV," Wikipedia, Wikimedia Foundation, last edited November 17, 2019. As of April 26, 2022:
https://en.wikipedia.org/wiki/Creaton_UAV

Delair, "Delair UX11: The Smartest Mapping Drone," webpage, undated. As of April 26, 2022:
https://delair.aero/delair-commercial-drones/professional-mapping-drone-delair-ux11/

Delta Drone, "Fixed-Wing UAV Delta Y," Direct Industry, undated. As of April 25, 2022:
https://www.directindustry.com/prod/delta-drone/product-100907-1788995.html

———, "Delta Y: UAV Professional System," brochure, c. 2017. As of April 25, 2022:
https://www.deltadrone.com/wp-content/uploads/2017/04/dd_delta_y_brochure_en.pdf

DJI, "Agras T10: The Ideal Drone for New Farmers," webpage, undated a. As of May 9, 2022:
https://www.dji.com/t10

———, "Agras T30: A New Flagship for Digital Architecture," webpage, undated b. As of May 9, 2022:
https://www.dji.com/t30

Dragonfly Pictures, "DP-12 Rhino," webpage, undated. As of April 25, 2022:
https://www.dragonflypictures.com/products/dp-12-rhino/

Edaran Mesra, "Unmanned Arial Vehicle (Fuji Imvac, Japan)," webpage, undated. As of May 9, 2022:
https://www.edaranmesra.com/product-Fuji-Imvac.html

EHang, "EHang's Smart Logistics Ecosystem," webpage, undated. As of April 26, 2022:
https://www.ehang.com/logistics/

Elektra Solar, "Elekra Two Solar," webpage, undated. As of May 9, 2022:
https://www.elektra-solar.com/products/elektra-two-solar/

ElevonX, untitled leaflet, c. October 2020. As of May 9, 2022:
https://www.elevonx.com/wp-content/uploads/2020/10/ElevonX_Leaflet.pdf

e.Yo, homepage, undated. As of April 26, 2022:
http://www.eyo-copter.com/

Flying Production, "Da-Vinci VTOL sUAS," webpage, undated. As of April 25, 2022:
https://www.flying-production.com/da-vinci

FlyingCinema, "CineTank MK 2," product listing, undated.

Hann, Richard, and Joachim Wallisch, "UAV Database," version 1.1, DataverseNO, 2020. As of December 2021:
https://dataverse.no/dataset.xhtml?persistentId=doi:10.18710/L41IGQ

Harpia Tech, homepage, undated. As of May 9, 2022:
https://harpia-tech.com/

Harris Aerial, "Carrier H6 Hybrid," webpage, undated. As of April 25, 2022:
https://www.harrisaerial.com/carrier-h6-hybrid-drone/

Helicam, homepage, undated. As of April 26, 2022:
https://www.helicam.asia/

Helipse, "HE190ES: Versatile, Agile, Efficient," webpage, undated. As of April 26, 2022:
https://helipse.com/en/product/HE190

Hexo+, homepage, undated. As of April 26, 2022:
https://hexoplus.com/

HighEye, homepage, undated. As of April 25, 2022:
https://www.higheye.com/

HobbyKing, "H-King Bixler 3 (PNF) Glider 1550mm (61")," product listing, undated. As of April 25, 2022:
https://hobbyking.com/en_us/h-king-bixler-3-glider-1550-pnf.html?___store=en_us

Hoffer, Nathan V., *System Identification of a Small Low-Cost Unmanned Aerial Vehicle Using Flight Data from Low-Cost Sensors*, master's thesis, Logan, Utah: Utah State University, May 2014. As of April 26, 2022:
https://digitalcommons.usu.edu/cgi/viewcontent.cgi?article=5289&context=etd

IDETEC Unmanned Systems, "Remarkable," webpage, undated. As of May 9, 2022:
https://www.stardust-uas.com/specifications

INAV, "SISUAR: IMINT/UAV Surveillance and Reconnaissance System," webpage, undated. As of May 9, 2022, in Romanian:
https://www.inav.ro/securitate_sisuar.php

Inkonova, "TILT Ranger," webpage, undated.

Integrated Dynamics, "Rover MK II XS," brochure, c. June 2019a. As of May 9, 2022:
https://idaerospace.com/wp-content/uploads/2019/06/Rover-XS.pdf

———, "Solaris," brochure, c. November 2019b. As of May 9, 2022:
https://idaerospace.com/wp-content/uploads/2019/11/SOLARIS-PDF.pdf

Janes Markets Forecast, "Unmanned Air Vehicles," undated. Downloaded November 18, 2021.

jDrones, untitled product page, undated. As of April 26, 2022:
https://www.jdrones.com/products-landing.html

Joint Chiefs of Staff, *Joint Concept of Operations for Unmanned Aircraft Systems*, version 1.5, undated, Not available to the general public.

La Franchi, Peter, "Mexico Throws Its Hat into the Ring with Ehecatl," FlightGlobal, June 17, 2007. As of May 20, 2022:
https://www.flightglobal.com/mexico-throws-its-hat-into-the-ring-with-ehecatl/74391.article

Lozano Rocabeyera, Ferran, *Study of a Feasible Solution for a Specific Mission with Unmanned Aerial Vehicles (UAV/RPAS)*, Escola Superior d'Enginyeries Industrial, Aeroespacial i Audiovisual de Terrassa, Universitat Politècnica de Catalunya, BarcelonaTech, June 2016. As of April 26, 2022:
https://upcommons.upc.edu/bitstream/handle/2117/100782/ANNEX_108.pdf

MacDonald, James M., Penni Korb, and Robert A. Hoppe, *Farm Size and the Organization of U.S. Crop Farming*, U.S. Department of Agriculture, Economic Research Service, Economic Research Report 152, August 2013. As of June 8, 2022:
https://agris.fao.org/agris-search/search.do?recordID=US2022208674

Malloy Aeronautics, "T80 | TRV80," webpage, undated. As of April 26, 2022:
https://www.malloyaeronautics.com/t80/

Matthew, Kyle, "RitewingRC Zephyr II Review—with Video!" *Model Airplane News*, July 16, 2015. As of April 26, 2022:
https://www.modelairplanenews.com/ritewingrc-zephyr-ii-review/

Mavtech, "Q4E Drone: Overview," webpage, undated. As of April 26, 2022:
https://www.mavtech.eu/en/products/q4e-drone/

microUAV, "Scythe," webpage, undated. As of April 26, 2022:
http://www.microuav.com/AirVehicle/Scythe

Mikado Model Helicopters, homepage, undated. As of April 26, 2022:
http://www.mikado-heli.de/

Mjaawad, "Chinese ASN-209 Unmanned Aerial Vehicle (UAV)," *Chinese Military Review*, October 20, 2011. As of May 9, 2022:
https://chinesemilitaryreview.blogspot.com/2011/10/chinese-asn-209-tactical-unmanned_20.html

Mobile Recon Systems, "Dauntless (Prototype)," webpage, undated. As of April 25, 2022:
https://mobilereconsystems.com/dauntless/

National Chung-Shan Institute of Science and Technology, "Cardinal II Unmanned Aircraft System," webpage, last updated July 10, 2020. As of May 9, 2022:
https://www.ncsist.org.tw/eng/csistdup/products/product.aspx?product_id=268&catalog=41

ntc_tech, "Aeromao Aeromapper EV2 Aircraft Drone," auction listing, eBay, last updated September 24, 2021. As of April 25, 2022:
https://www.ebay.com/itm/Aeromao-Aeromapper-EV2-Aircraft-Drone-/153390941953?_ul=IL

OM UAV Systems, "Curiosity Quadcopter UAV," webpage, undated. As of April 25, 2022:
http://www.omuavsystems.com/curiosity-quadcopter-uav.html

Parmar, Tekendra, "Drones in Southeast Asia," Center for the Study of the Drone at Bard College, August 14, 2015. As of May 9, 2022:
https://dronecenter.bard.edu/drones-in-southeast-asia/

Petrova, Magdalena, and Lora Kolodny, "Zipline's New Drone Can Deliver Medical Supplies at 79 Miles per Hour," *The Edge*, CNBC, April 3, 2018. As of April 26, 2022:
https://www.cnbc.com/2018/04/02/zipline-new-zip-2-drone-delivers-supplies-at-79-mph.html

"PR-5 Wiewior Plus," *Wikipedia*, Wikimedia Foundation, last edited January 30, 2022. As of April 26, 2022:
https://en.wikipedia.org/wiki/PR-5_Wiewior_plus

Prodrone, "PD8X," webpage, undated. As of May 9, 2022:
https://www.prodrone.com/products/pd8x/

Public Law 107-296, Homeland Security Act of 2002, November 25, 2002. As of May 12, 2019:
https://www.govinfo.gov/app/details/PLAW-107publ296

"PUT UAV," *Wikipedia*, Wikimedia Foundation, last edited June 20, 2020. As of April 26, 2022:
https://en.wikipedia.org/wiki/PUT_UAV

"Reiner Stemme Q01," *Military Factory*, last edited October 30, 2020. As of May 9, 2022:
https://www.militaryfactory.com/aircraft/detail.php?aircraft_id=1621

Repreneurs.com, "Jugement de conversion en liquidation judiciaire: AIRBORNE CONCEPT située à Cugnaux (31270) a été déclarée en Jugement de conversion en liquidation judiciaire par le TRIBUNAL DE COMMERCE DE TOULOUSE" [Judgment of conversion into judicial liquidation: Airborne Concept located in Cugnaux (31270) has been declared in judgment of conversion into judicial liquidation by the Toulouse Commercial Court], judgment issued September 27, 2018; record updated February 9, 2021. As of April 26, 2022, in French:
https://www.repreneurs.com/jugement-de-conversion-en-liquidation-judiciaire/323471-airborne-concept

Rotomation, homepage, undated.

Safran, "Patroller™: Long-Endurance, Multi-Mission and Multi-Sensor Tactical UAV System," webpage, undated. As of May 9, 2022:
https://www.safran-group.com/products-services/patrollertm-long-endurance-multi-mission-and-multi-sensor-tactical-uav-system

Samoloty w Lotnictwie Polskim [Airplanes in Polish aviation], "PR-5 'Wiewiór,' 2009," webpage, undated. As of April 26, 2022, in Polish:
http://www.samolotypolskie.pl/samoloty/2153/126/PR-5-Wiewior

Shandong Black Shark Intelligent Technology, homepage, undated.

Shenzhen Joyton Innovation Technology, "Agricultural Drone: Oxygen Precision Agriculture," webpage, undated. As of May 9, 2022:
http://www.joyton.cn/en/tianyou1.html

Shenzhen Keweitai Enterprise Development, "KWT-Z4M-H," webpage, undated. As of April 26, 2022:
http://en.keweitai.com/kwtz4mh

Shenzhen Xiangnong Innovation Technology, "R-16," webpage, undated. As of April 26, 2022:
http://www.txauav.com/en/html/en/r16.html

"SKNL PRz PR-5 Wiewior" [Технические характеристики. Фото Подробнее на], blog post, Avia.pro, June 9, 2016. As of April 26, 2022, in Russian:
https://avia.pro/blog/sknl-prz-pr-5-wiewior-tehnicheskie-harakteristiki-foto

Sky Eye, "Micro," webpage, undated. As of April 26, 2022:
https://skyeyeinnovations.se/start/airborne-solutions/sky-eye-micro/

S&T—See Science and Technology Directorate.

SteelRock Technologies, "UAV Platforms," brochure, c. 2019. As of April 26, 2022:
https://www.acxodus.com/wp-content/uploads/2019/03/SRT-UAV-Brochure-19.09.18.pdf

"T23E UAV: Eleron-3," Milipedia, c. 2012. As of April 26, 2022, in Polish:
https://www.infolotnicze.pl/2012/05/29/t23e-uav-eleron-3/

Teal Drones, "Teal One," webpage, undated. As of April 26, 2022:
https://tealdrones.com/teal-one/

Teledyne FLIR, "Advanced Multi-Mission UAS: SkyRanger® R70," webpage, undated. As of May 9, 2022:
https://www.flir.com/products/skyranger-r70/

Thunder Tiger, homepage, undated.

Tianyu Chuangtong Technology [北京天宇创通科技有限公司 版权所有], homepage, undated. As of April 26, 2022, in Chinese:
http://www.uav-china.com/

Topcon, Topcon Sirius: Unmanned Aerial Solution, brochure, rev. B, August 2015. As of April 26, 2022:
https://www.topconsolutions.com/images/topcon/brochure/sirius_solutions_catalog.pdf

Trancomm Technologies [版权所有], homepage, undated. As of April 26, 2022, in Chinese:
http://www.trancomm.com.cn/index.html

UASUSA, "The Recon," webpage, undated.

UAV Research Lab, "Rotor Buzz," webpage, undated. As of April 26, 2022:
http://uavrl.com/rotor_buzz.html

UAVision, "WINGO S," webpage, undated. As of April 26, 2022:
https://uavision.wixsite.com/uavision/wingo-s

UAVOS, homepage, undated. As of April 26, 2022:
https://www.uavos.com/

UCONSYSTEM, "Fixed-Wing UAV REMOEYE-002B," AeroExpo, undated a. As of April 26, 2022:
https://www.directindustry.com/prod/uconsystem/product-101655-2220773.html

———, "Industrial Use," webpage, undated b. As of April 26, 2022:
http://www.uconsystem.com/eng/products/industrial/remom-002.asp

———, "Professional UAV: REMO M-002," AeroExpo, undated c. As of April 26, 2022:
https://www.aeroexpo.online/prod/uconsystem-co-ltd/product-175105-35320.html

Unmanned Systems and Solutions, "Long Endurance Aerial Platform," webpage, undated. As of April 26, 2022:
https://www.unmannedsas.com/leap

U.S. Code, Title 6, Domestic Security; Chapter 1, Homeland Security Organization; Subchapter III, Science and Technology in Support of Homeland Security; Section 185, Federally Funded Research and Development Centers. As of March 20, 2021:
https://uscode.house.gov/view.xhtml?req=(title:6%20section:185%20edition:prelim)

Vanilla Unmanned, homepage, undated. As of April 26, 2022:
https://vanillaunmanned.com/

Volacom, "Volacopter LZ73-PRO UAV," webpage, undated. As of May 9, 2022:
https://www.volacom.com/solutions/volacopter-lz73-pro-uav/

VulcanUAV, "Aircraft," webpage, undated. As of April 26, 2022:
https://vulcanuav.com/aircraft/

Wilson, Bradley, Shane Tierney, Brendan Toland, Rachel M. Burns, Colby P. Steiner, Christopher Scott Adams, Michael Nixon, Raza Khan, Michelle D. Ziegler, Jan Osburg, and Ike Chang, *Small Unmanned Aerial System Adversary Capabilities*, Homeland Security Operational Analysis Center operated by the RAND Corporation, RR-3023-DHS, 2020. As of February 15, 2022:
https://www.rand.org/pubs/research_reports/RR3023.html

XAG, "XAG P40 Agricultural Drone," webpage, undated a. As of May 9, 2022:
https://www.xa.com/en/p40

———, "XAG V40 Agricultural Drone," webpage, undated b. As of May 9, 2022:
https://www.xa.com/en/v40

Yamaha Motor Corporation, homepage, undated. As of April 26, 2022:
https://yamaha-motor.com/

Yuneec Holding, "Typhoon H Plus," webpage, undated. As of May 9, 2022:
https://us.yuneec.com/typhoon-h-plus/

Zala Aero Group, homepage, undated a.

———, "UAV ZALA 421-06" [БПЛА ZALA 421-06], webpage, undated b.

———, "Unmanned aerial vehicle ZALA 421-16E2" [Беспилотное воздушное судно ZALA 421-16E2], webpage, undated c.

"Zipline (Drone Delivery)," *Wikipedia*, Wikimedia Foundation, last edited April 26, 2022. As of April 26, 2022:
https://en.wikipedia.org/wiki/Zipline_(drone_delivery)